The Superiority of Line Breeding
Over Narrow Breeding

by O.F. Cook

with an introduction by Jackson Chambers

Self Reliance Books

Get more historic titles on animal and stock breeding, gardening and old fashioned skills by visiting us at:

http://selfreliancebooks.blogspot.com/

Introduction

I am pleased to present yet another title on the Principles of Animal Breeding.

This volume is entitled "The Superiority of Line Breeding Over Narrow Breeding" and was published in 1909.

The work is in the Public Domain and is re-printed here in accordance with Federal Laws.

As with all reprinted books of this age that are intended to perfectly reproduce the original edition, considerable pains and effort had to be undertaken to correct fading and sometimes outright damage to existing proofs of this title. At times, this task is quite monumental, requiring an almost total "rebuilding" of some pages from digital proofs of multiple copies. Despite this, imperfections still sometimes exist in the final proof and may detract from the visual appearance of the text.

I hope you enjoy reading this book as much as I enjoyed making it available to readers again.

Jackson Chambers

BUREAU OF PLANT INDUSTRY.

Physiologist and Pathologist, and Chief of Bureau, Beverly T. Galloway.
Physiologist and Pathologist, and Assistant Chief of Bureau, Albert F. Woods.
Laboratory of Plant Pathology, Erwin F. Smith, Pathologist in Charge.
Fruit Disease Investigations, Merton B. Waite, Pathologist in Charge.
Investigations in Forest Pathology, Haven Metcalf, Pathologist in Charge.
Cotton and Truck Diseases and Plant Disease Survey, William A. Orton, Pathologist in Charge.
Pathological Collections and Inspection Work, Flora W. Patterson, Mycologist in Charge.
Plant Life History Investigations, Walter T. Swingle, Physiologist in Charge.
Cotton Breeding Investigations, Archibald D. Shamel and Daniel N. Shoemaker, Physiologists in Charge.
Tobacco Investigations, Archibald D. Shamel, Wightman W. Garner, and Ernest H. Mathewson, in Charge.
Corn Investigations, Charles P. Hartley, Physiologist in Charge.
Alkali and Drought Resistant Plant Breeding Investigations, Thomas H. Kearney, Physiologist in Charge.
Soil Bacteriology and Water Purification Investigations, Karl F. Kellerman, Physiologist in Charge.
Bionomic Investigations of Tropical and Subtropical Plants, Orator F. Cook, Bionomist in Charge.
Drug and Poisonous Plant and Tea Culture Investigations, Rodney H. True, Physiologist in Charge.
Physical Laboratory, Lyman J. Briggs, Physicist in Charge.
Agricultural Technology, Nathan A. Cobb, Crop Technologist in Charge.
Taxonomic and Range Investigations, Frederick V. Coville, Botanist in Charge.
Farm Management, William J. Spillman, Agriculturist in Charge.
Grain Investigations, Mark Alfred Carleton, Cerealist in Charge.
Arlington Experimental Farm and Horticultural Investigations, Lee C. Corbett, Horticulturist in Charge.
Vegetable Testing Gardens, William W. Tracy, sr., Superintendent.
Sugar-Beet Investigations, Charles O. Townsend, Pathologist in Charge.
Western Agricultural Extension, Carl S. Scofield, Agriculturist in Charge.
Dry-Land Agriculture Investigations, E. Channing Chilcott, Agriculturist in Charge.
Pomological Collections, Gustavus B. Brackett, Pomologist in Charge.
Field Investigations in Pomology, William A. Taylor and G. Harold Powell, Pomologists in Charge.
Experimental Gardens and Grounds, Edward M. Byrnes, Superintendent.
Foreign Seed and Plant Introduction, David Fairchild, Agricultural Explorer in Charge.
Forage Crop Investigations, Charles V. Piper, Agrostologist in Charge.
Seed Laboratory, Edgar Brown, Botanist in Charge.
Grain Standardization, John D. Shanahan, Crop Technologist in Charge.
Subtropical Garden, Miami, Fla., P. J. Wester, in Charge.
Plant Introduction Garden, Chico, Cal., W. W. Tracy, jr., Assistant Botanist in Charge.
South Texas Garden, Brownsville, Tex., Edward C. Green, Pomologist in Charge.
Farmers' Cooperative Demonstration Work, Seaman A. Knapp, Special Agent in Charge.
Seed Distribution (Directed by Chief of Bureau), Lisle Morrison, Assistant in General Charge.

Editor, J. E. Rockwell.
Chief Clerk, James E. Jones.

BIONOMIC INVESTIGATIONS OF TROPICAL AND SUBTROPICAL PLANTS.

SCIENTIFIC STAFF.

O. F. COOK, *Bionomist in Charge.*

G. N. Collins and F. L. Lewton, *Assistant Botanists.*
H. Pittier, J. H. Kinsler, and A. McLachlan, *Special Agents.*
C. B. Doyle and R. M. Meade, *Scientific Assistants.*

146
2

LETTER OF TRANSMITTAL.

U. S. Department of Agriculture,
Bureau of Plant Industry,
Office of Chief of Bureau,
Washington, D. C., February 25, 1909.

Sir: I have the honor to transmit herewith a paper entitled " The Superiority of Line Breeding over Narrow Breeding," by Mr. O. F. Cook, and recommend its publication as Bulletin No. 146 of the special series of the Bureau of Plant Industry.

Respectfully,

B. T. Galloway,
Chief of Bureau.

Hon. James Wilson,
Secretary of Agriculture.

CONTENTS.

B. P. I.—444.

THE SUPERIORITY OF LINE BREEDING OVER NARROW BREEDING.

INTRODUCTION.

None of the applications of the science of evolution to the art of breeding has been the subject of so much study, experiment, and discussion as the relative merits of inbreeding and cross-breeding. Nevertheless, opinions remain as discordant as ever, for some experiments appear to indicate that cross-breeding is better than inbreeding, while other experiments seem to show quite as definitely that inbreeding is better than crossing.

Experiments are our means of securing answers to scientific questions. When experiments appear to give contradictory or equivocal answers we know that we have not asked our questions in the right way. We are warned that there are more differences among our facts than our theories have recognized. We must find another point of view more favorable for the interpretation of the facts.

Some breeders have generalized upon the results of their own experiments and ignored the contrary results reported elsewhere. Others have taken a broader view of the subject and have sought to compromise the issue between inbreeding and cross-breeding by assuming that the different types of plants or animals are so differently constituted that inbreeding represents the normal method for some species, cross-breeding for others. The effect of this opinion has been merely to postpone the issue, not to determine it, for we can not avoid thinking of reproduction as a physiological process or fail to ask the question why members of one species should appear to thrive best by inbreeding and members of another species by crossing.

It is the purpose of this paper to point out some of the sources of confusion which have interfered with definite scientific solutions of problems of breeding, and to indicate a point of view from which contradictory opinions can be reconciled. As soon as it becomes possible to recognize physiological relations behind the apparently contradictory evidence the subject is opened to new methods of investigation.

THREE PRINCIPAL TYPES OF REPRODUCTION.

I. Broad breeding.
II. Narrow breeding.
III. Line breeding:
    (*a*) In-and-in breeding.
    (*b*) Self-fertilization.
    (*c*) Parthenogenesis.
    (*d*) Vegetative propagation.

Some of the difficulties in the study of breeding have come from the fact that three principal types or methods of reproduction have been under consideration, though only two have been formally recognized. Many writers on breeding could be charged with the error of reasoning called by logicians "the undistributed middle." They have applied the same names to conditions essentially different.

The term "cross-breeding" has been applied to the condition of free interbreeding among the members of large groups, such as species in nature, as well as to the mating of individuals representing small, close-bred varieties, or even to the crossing of closely related individuals of the same strain when such crosses are being contrasted with self-fertilization.

The term "inbreeding" has also been used for a wide series of conditions. It is commonly applied to the self-fertilization of plants by their own pollen as well as to in-and-in breeding among closely related members of the same family group, and even to the more miscellaneous form of close breeding found in a flock of sheep or poultry when no "new blood" is brought in.

The meanings of the two terms remain distinct only at the extreme ends of the series of facts to which they are applied, and widely overlap in the middle. Whether a particular condition of descent is to be described as inbreeding or as cross-breeding is usually allowed to depend entirely on the particular instances in hand. The same case may be called "inbreeding" if the author wishes to compare it with another in which there is wider crossing, or it may be called "cross-breeding" if it is to be compared with a smaller amount of crossing. Thus the same word may describe different conditions, or different words the same condition.

For many purposes of comparison the use of the words "cross-breeding" and "inbreeding" in these merely relative senses is convenient and entirely proper, but it leads to serious confusion when we attempt to distinguish the underlying factors. This confusion can be avoided by choosing other words to which more definite meanings can be attached. For the purposes of the present explanation it seems desirable to recognize three principal conditions or methods of reproduction, which may be characterized as follows:

Broad breeding is the condition of descent found in natural species, which consist of millions of diverse individuals freely interbreeding with each other, so that the vast numbers of lines of descent of the species are joined into a broad network.[a]

Narrow breeding is the condition of descent found in carefully selected varieties, consisting of relatively small numbers of closely similar individuals interbreeding with each other to form a narrow network of descent.

Line breeding is the condition of descent found in strains descended from single individuals propagated without interbreeding with other lines of descent, so that no network is formed.

The comparison of descent in normal broad-bred species to a network or fabric helps to distinguish between the two types of restricted descent. The usual process of selection to secure uniformity in the expression of characters in a sexually reproduced group is called "narrow breeding" because it reduces or subdivides the network of the species to narrow strands. When descent is still further restricted so that lines of descent from particular individuals are kept separate we have the condition called "line breeding."

Four forms of line breeding are to be considered in later sections: Vegetative propagation, parthenogenesis, self-fertilization, and in-and-in breeding.

Broad breeding might be described as a wide and indiscriminate form of cross-breeding, and line breeding as the strictest form of inbreeding. Narrow breeding is the intermediate condition which many authors have called "cross-breeding" and many others "inbreeding." The mistake has been to suppose that this intermediate condition of narrow breeding is the same as one or the other of the extremes of the series. Narrow breeding is in reality quite as different from broad breeding and from line breeding as these are from each other, and in some very important respects more different, as will be shown in the later sections.

EFFECTS OF RESTRICTION OF DESCENT.

Broad breeding is the normal condition of reproduction in natural species, the condition in which normal evolution takes place. The lines of descent of any individual if traced backward are found to diverge more and more to connect with a vast number of ancestors, and eventually with all the members of the group, for the network is continuous. It is this continuous network of descent that makes a species a definite biological entity and not a mere collection of

[a] Cook, O. F. The Vital Fabric of Descent. Proceedings, Washington Academy of Sciences, vol. 7, pp. 301–323. 1906.

similar individuals. The absence of any restriction of interbreeding among the members of a species allows the lines of descent to be joined into a broad network. Restriction of descent means a limitation of the number of ancestors and of lines of descent, so that only a narrow network, or none at all, can be formed.

If we recognize broad breeding as the natural condition in which organisms thrive and species make evolutionary progress, it is easy to understand that persistent narrow breeding may interfere with such progress or cause degeneration in characters or qualities already attained. Evolution, instead of being brought about by restriction of descent, as some have supposed, is actually hindered and finally undone.[a]

There is no evidence to show that any form of restricted descent, either to narrow networks or to simple individual lines, is an advantage to a group of organisms in the sense that it enables them to produce stronger individuals in larger numbers. The evidence all tends to show that restriction of descent results, sooner or later, in degeneration and extinction. Nevertheless, different methods of restricting descent may bring very different results, which the distinction between narrow breeding and line breeding helps us to recognize. If biological phenomena were governed by rules of logic it could be reasoned that line breeding must be more promptly injurious than narrow breeding, because it is narrower than narrow breeding. The actual results often show the contrary. One line of descent often proves to be stronger than a few lines. Line breeding by self-fertilization or in-and-in breeding is often superior to more indiscriminate breeding among the descendants of a few ancestors. After distinguishing the three general types or conditions of reproduction, it becomes possible to consider that line breeding is often superior to narrow breeding, without any need of supposing that line breeding is ever superior to broad breeding, in maintaining higher degrees of vigor and fertility.

Vigor and fertility, though of fundamental biological and evolutionary importance, are not the only forms of superiority that the breeder takes into account, for many highly prized characteristics of domesticated varieties, such as seedlessness, are in the nature of biological degenerations artificially preserved for the special uses of man. We think of even a seedless variety as fertile if it yields us good quantities of a useful product, though it has to be reckoned as completely sterile in the biological sense. But whether we wish in

---

[a] Cook, O. F. Methods and Causes of Evolution. Bulletin 136, Bureau of Plant Industry, U. S. Department of Agriculture. 1908. See also Aspects of Kinetic Evolution, in Proceedings, Washington Academy of Sciences, vol. 8, p. 197, 1907.

particular cases to preserve the biological superiority of our varieties or are willing to sacrifice vigor or fertility to special purposes of use, we need to recognize the underlying biological principles. We can then understand that different kinds of improvement are to be sought in different directions and by different methods.

AGRICULTURAL IMPROVEMENT BY INCREASE OF UNIFORMITY.

The object of restricting descent to superior individuals is to secure progeny as close as possible to the parental standard of excellence. This object is best attained when the progeny continue to follow each other in an unbroken series of individuals closely similar to the high-grade ancestor and closely alike among themselves.

Increased uniformity often constitutes an agricultural improvement of a variety quite apart from the question whether any individuals of the variety advance beyond the standard of the carefully selected ancestor of the line-bred group. It has often been supposed that progeny obtained under restriction of descent are actually superior to any of their broad-bred ancestors, but this is now seriously questioned, and many recent writers have denied it altogether. The chief advantage gained through restriction of descent is conservative rather than constructive. The practical improvement of varieties by selection and other forms of restricted descent lies in the preservation of characters that already exist rather than in the attainment of new characters.

VEGETATIVE PROPAGATION THE MOST EFFECTIVE METHOD OF LINE BREEDING.

There can be no doubt that the most effective way of securing the desired uniformity and of maintaining it for long periods of time is by the process of vegetative propagation as applied to many cultivated plants. To prove this by new experiments might require far longer than the lifetime of a man, but the history of agriculture makes such experiments unnecessary. We know that many tropical root crops and other plants have been propagated from cuttings since very remote periods, to be counted in thousands of years. Our carefully selected narrow-bred varieties of seed-propagated plants have been known for only a few decades at the most. The majority of them go out of use and disappear after only a few generations, giving place to other " new " varieties, better at first than the old, but giving no more assurance of permanent superiority.

VEGETATIVE PROPAGATION DEPENDS ON LONGEVITY.

To treat vegetative propagation as a form of line breeding may appear unwarranted in view of the current opinion that vegetative propagation is a purely nonsexual process, whereas self-fertilization

146

and in-and-in breeding are usually viewed as forms of sexuality. It is true that vegetative propagation is often carried on by parts of plants which have no apparent connection with the organs of sexual reproduction, but there are other and more fundamental considerations which show that all of our higher types of plants are continuously dependent for their existence upon the sexual process of conjugation, without regard to whether particular individuals are raised from seed or grown from cuttings.[a]

It is misleading to suppose that vegetative propagation as carried on by the higher plants involves a cessation of conjugation. New conjugations become unnecessary ·because the old conjugation is greatly prolonged. Thus vegetative propagation is not to be considered as a substitute for conjugation, but as evidence of an ability of the cells of the plants to continue in a state of conjugation without the need of frequent renewal. This power to maintain existence by vegetative propagation depends upon a special property or quality not shown in plants whose cells are able to remain in a state of conjugation only for the lifetime of a single seed-propagated individual. Vegetative propagation is to be viewed as a form of longevity of the protoplasm, enabling growth to continue without a new conjugation. Without this power of longevity, vegetative propagation would be as impossible as it would be for an annual herb to grow into a tree. Species and varieties differ greatly in their powers of vegetative propagation, just as they differ in other forms of longevity.

Many writers have recognized that conjugation is a means of rejuvenation or renewal of the energy of the protoplasm. The bodily activities of organisms appear greater in young plants and animals and lessen with age. The renewal of the energy of the protoplasm by conjugation is effective in different species for very different periods of time. Many plants are annuals, living for only a few weeks or months, while others have an enormous longevity. The ages of individual trees of several different species have been reckoned in hundreds or thousands of years. To maintain the life of such organisms the process of conjugation must have a long-sustained efficiency, avoiding the need of the frequent renewal, as in vegetative propagation.

The longevity of a plant capable of vegetative propagation is not to be measured by the life of a single individual, but by the length of the series of vegetative individuals that can be produced without renewed conjugation. Differences of longevity probably represent

---

[a] Cook, O. F., and Swingle, W. T. Evolution of Cellular Structures. Bulletin 81, Bureau of Plant Industry, U. S. Department of Agriculture. 1905.

differences in the structure or quality of the protoplasm itself. Rudely, we may compare the organisms to clocks and say that some have stronger mainsprings or are more effectively wound up. In vegetative propagation we may say that the protoplasmic mainsprings outlast the individual organisms and remain serviceable through a long series of organisms.

Longevity is not proportional to size. The persistent vegetative growth of some of our herbaceous plants might enable them to excel the largest trees if all the tissues originating from a single seed could be kept together as in the case of the trees. The tree lives only as long as the protoplasm of the cells retains its vigor, or until it becomes too large to support or nourish itself, but with a perennial herb the longevity of the protoplasm is passed on through many vegetative generations. The individual plant may not remain alive for more than a year or two, and yet the same rootstock may creep along in the ground for decades or centuries. The greater longevity of plant organisms may depend somewhat upon the fact that they continue to form new tissues instead of attaining a definite maturity merely from the different habits of growth. The weakened vitality of the protoplasm may be the same in the vegetative variety as in a tree or a long-lived animal, the difference of visible results arising merely from the different habits of growth. The weakened vitality of old varieties of potatoes or of sugar cane may be compared with the gradual weakening of aged trees or of aged men. There is a slackening of the organic energies which can be quickened only by new conjugations.

In a species in which the individuals are short lived we think of a new conjugation as necessary to restore the energy of the protoplasm for the growth of each generation. With long-lived species or those capable of vegetative propagation it is evident that there is no such necessity of frequent renewal of conjugation.

Additional insight regarding the nature of conjugation has been gained in recent years. Instead of conjugation being a process which takes place only when a new generation is to begin, it is now known that the cells which compose the bodies of the higher plants and animals all represent a state of prolonged conjugation. They are not like the simple cells which are formed *between conjugations* in the lower and more primitive groups, but are double cells like those which are formed in the lower groups only *during conjugation*. The complex bodies of the higher plants and animals are built up, not merely because conjugation takes place, but because conjugation continues throughout the lifetime of the individual, however long this may be.

OTHER METHODS OF LINE BREEDING.

Having once recognized that the vegetative propagation of our cultivated plants is a form of longevity, it becomes easy to see that other methods of line breeding are very similar to vegetative propagation. The only difference between vegetative propagation and parthenogenesis is that in the latter the vegetative development of the new individual arises from tissues which usually serve the purpose of reproduction rather than from parts that are purely vegetative. The cells which normally produce the germ cells grow into a small embryo-like bud, so that seeds are developed without access of pollen.

Self-fertilization is not far from parthenogenesis, for it is accomplished by reuniting cells of the same plant only recently separated. In self-fertilization the appearance of normal conjugation is maintained, but the results do not correspond to those of the interbreeding of different individuals. Self-fertilized types show the same uniformity of character expression as vegetative varieties, the same failure to permanently maintain the life of the stock, and the same gradual loss of vitality with age.

In-and-in breeding is the nearest approach to self-fertilization possible in plants or animals which have the individuals of different sexes. The germ cells which unite are from the most closely related individuals.

In vegetative propagation and parthenogenesis there is no interruption of the process of conjugation, for no new germ cells are formed. In self-fertilization and in-and-in breeding germ cells are formed, but the conditions of normal sexual reproduction are avoided, for cells of the same ancestry are brought together again, instead of cells of different ancestry. Though the formalities of sexual reproduction are repeated, there is in reality no more sexuality, in the physiological sense, than in vegetative propagation, for the unions of cells represent mere renewals of old conjugations, instead of being truly normal new conjugations between partners representing different lines of descent.

Parthenogenesis is plainly intermediate between vegetative propagation and self-fertilization. Parthenogenetic and self-fertilized types can not be found, of course, in a state of complete seedlessness like some of our vegetative varieties, for seedlessness in a seed-propagated variety would mean immediate extinction. That so many vegetative varieties belonging to very diverse families have become seedless is one of many evidences that even the most successful form of line breeding does not maintain the full and normal vigor of organic types. The frequency of vegetative propagation and self-fertilization among cultivated species has tended to give a misleading idea of the general importance of these methods of reproduction.

A few species of plants, and especially certain degenerate parasites, appear to be uniformly parthenogenetic, but in the great majority of cases parthenogenesis appears as an alternative of normal fertilization, just as many species resort to self-fertilization or to in-and-in breeding when there are no opportunities of normal crossing.

The domestication of perennial plants propagated by cuttings was very much easier for primitive man than that of annual species of which seed had to be saved. Thus the earliest domestications were confined largely to perennial root crops and to trees, the annual species being relatively late acquisitions, as shown by the fact that most of them are still known in the wild state.[a]

Many of our self-fertilized domesticated types have open-fertilized wild relatives, indicating that the habit of self-fertilization was not the primitive condition of descent. The same is true of most, if not all, of the self-fertilized wild plants; either they are occasionally crossed or they have cross-fertilizing relatives. Among the seed-propagated cultivated species the tendency of selection is generally toward self-fertilization, even where no such selection has been intended.

Plants are often carried into regions where their insect friends or other natural agents of cross-fertilization are absent or where the climatic conditions are unfavorable for the transfer of pollen through the air. The varieties of wheat and other cereals developed in northern Europe show more pronounced adaptations for self-fertilization than those of the Mediterranean countries. This can be explained by the selective action of the northern climate. Unfavorable weather at the time of flowering might cause a total failure in a crop dependent upon cross-pollination and allow seed to be saved only from self-pollinated plants. In a similar way it is possible to explain the special prevalence of vegetative propagation, parthenogenesis, and self-fertilization among plants that bloom early in the spring when unfavorable weather is likely to prevent cross-pollination by insects.

RELATION OF SELF-FERTILITY TO VEGETATIVE PROPAGATION.

Darwin found that some plants yielded better progeny by self-fertilization than when the pollen came from other individuals of the same stock, and later experimenters have reported similar results in several self-fertilizing species. Some of the most notable instances are reported among the varieties of tobacco studied by Mr. A. D.

---

[a] Cook, O. F. The American Origin of Agriculture. Popular Science Monthly, October, 1902. Reprinted in the Annual Report of the Smithsonian Institution for 1903, p. 481, under the title "Food Plants of Ancient America."

Shamel, of the Bureau of Plant Industry. Plants fertilized by their own pollen gave, in numerous instances, progeny much better on the average than plants fertilized by pollen from other individuals of the same stock. Individual members of the cross-fertilized series were equal to members of the self-fertilized series, but many were notably inferior as well as more diverse among themselves, which seriously detracts from the commercial value of the product.

These experiments are of special interest because it was ascertained at the same time that hybrids between two distinct strains of tobacco showed less individual diversity than when crosses were made among members of the same strain. Thus it becomes apparent that the diversity and relative inferiority of the progeny of the individual crosses is not to be ascribed to the crossing as such, but to the condition of narrow breeding which such a cross represents, as compared with the line-bred stock on the one hand and with the broad-bred crosses between strains on the other. We may think of the inferiority shown in the narrow-bred stocks as a result of insufficient diversity of descent as compared with broad-bred stocks, and at the same time we may recognize that self-fertilization serves to postpone the degeneration by combining cells as closely alike as possible.

Thus the analogies of vegetative propagation help us to understand how two cells of the closest relationship may continue reproduction to better advantage than those of slightly more distant relationship. Slight differences in the germ cells may be large enough to call forth individual diversity in the offspring and yet not large enough to give them the advantage of renewed vigor like that obtained by normal conjugation between germ cells derived from distinct lines of descent. The classing of self-fertilization with vegetative propagation is thus to be justified by the most practical reasons—that both attain the same results in producing a uniform progeny by eliminating the individual differences found among organisms produced by normal cross-fertilization.

That self-fertilization and cross-fertilization yield different results, the one of uniformity, the other of diversity, is a reason for believing that they constitute different processes of reproduction, though the nature of the difference is not yet known. The best suggestion of what this difference may be is afforded by a fact recently reported by Dr. Reginald R. Gates, of Chicago University, who has made a very detailed investigation of the processes of reproduction in a variety of evening primrose (*Oenothera rubrinervis*). The nuclei of the pollen mother cells were found to pass through a stage of contraction (synapsis) and to divide into pollen cells with the usual reduced number of chromosomes, but without giving any indication that a fusion takes place between two parallel strands of chromatin (mitapsis), the concluding act of the normal process of conjugation, described by many investigators of other plants and animals. Such a method of reproduction without a conjugation of chromatin (amitapsis) is very interesting as indicating that an apparently essential part of the process of conjugation may be omitted though all the external formalities of conjugation are preserved. Nevertheless, amitapsis may be considered a less violent departure from normal sexual processes than is parthenogenesis. Amitapsis could be reckoned as a production of sex cells by vegetative subdivision of a mother cell, parthenogenesis as the vegetative development of an undivided mother cell into a new organism.

If these observations on Oenothera should be confirmed and extended to other line-bred varieties we would have a reason for looking upon the individual diversity of cross-fertilized types as a result of the process of mitapsis, which appears to be omitted in these self-fertilized primroses.[a] And even if it were found that the behavior of the chromatin has no relation to other characteristics of the plants, we should not forget that uniformity and diversity are concrete facts in need of physiological explanation.

IN-AND-IN BREEDING.

Not a few plants are like the higher animals in having the sexes represented by separate individuals. In all such cases self-fertilization is of course an impossibility. The nearest approach to it is the mating of the most closely related individuals, sometimes called in-and-in breeding. Breeders of animals often secure better results by mating the most closely related individuals than by mating those that are a little less closely related, just as breeders of plants find the complete self-fertilization of a flower by pollen from its own stamens better than narrow crossing with pollen from other flowers of the same plant or from a closely related plant.

In some varieties breeding with close consanguinity has been applied with good results for a considerable series of generations. Other varieties have appeared to improve by in-and-in breeding, but soon show sterility or other degenerative weakness. In still other cases the reports indicate prompt and definite injury from in-and-in breeding. Relying on these individual experiences some breeders advocate

---

[a] "No indication of a doubling or pairing of the threads during these intermediate contraction stages could be observed, though they were carefully searched for. Moreover, in the earliest stages of the synaptic ball the thread appears to be as thin and delicate as in the reticulum, which does not favor the view that a pairing has taken place. The evidence, then, so far as it goes, is decidedly not in favor of a pairing."—R. R. Gates, A Study of Reduction in Oenothera Rubrinervis, Botanical Gazette, vol. 46, p. 8, July, 1908.

in-and-in breeding as a general principle of the art, while others condemn it as a wholly mistaken policy.

A part of the diversity of opinion is due, no doubt, to inherent differences in the longevity of the stocks. A part may also arise from the fact that some breeders have been dealing with poultry or other small animals which are usually allowed to run in flocks instead of being separated into individual lines of descent. While a flock might promptly decline under narrow breeding, careful in-and-in breeding of select individual lines of the same stock might preserve superior strains. Yet this same policy applied in a different species or even in a different variety of the same species might only hasten disaster, if the necessary protoplasmic longevity were lacking.

There is even less probability that a type can be permanently maintained by in-and-in breeding than by self-fertilization or by vegetative propagation, but the period of endurance is undoubtedly long enough in some species to give the method practical importance where uniformity and the special development of particular characters are more essential than to maintain the reproductive energy of the stock, the function in which the earliest decline may be expected. Thus it might be good policy to apply methods of strict self-fertilization or in-and-in breeding in localized varieties of cotton yielding special grades of fiber or in varieties of sheep producing a high-priced wool. The same systems might prove very unwise if the primary objects were to increase the total yield of cotton or to render the sheep more hardy and prolific.

SELF-FERTILIZATION SUPPLEMENTS BROAD BREEDING IN NATURAL SPECIES.

The flowers of some plants are so constructed that the pollen can readily fall upon the stigmas, while in others the parts are arranged so as to prevent self-fertilization. This fact has often been used to support the idea of a natural duality of methods of reproduction, an arbitrary difference of reproductive methods among the different species, not to be reconciled and interpreted by the same physiological principles. The failure to distinguish properly between the three different types of reproduction—broad breeding, narrow breeding, and line breeding—has kept contradictory opinions alive in the same way as in the question between cross-breeding and inbreeding.

The factor of longevity, the ability of the protoplasm to continue growth without a truly sexual reproduction, needs to be taken into account in understanding the great diversities of plants with respect to self-fertilization. Similar and nearly related types often differ widely in their ability to sustain themselves through self-fertilization, just as other groups of species and varieties differ among themselves in their powers of vegetative propagation.

If normal seeds can not be developed by self-fertilization, it is an advantage to a plant not to have its pollen fall on its own stigmas but to leave these unfertilized as long as possible, to increase the chances of arrival of the necessary foreign pollen. When good seeds can be formed by self-fertilization it becomes an advantage to the species to have its flowers so arranged that the stigmas are not left without pollen, just as it is an advantage to a plant able to propagate by vegetative growth to have also the habit of forming offsets. Whenever this power of sustained vitality exists it is obviously advantageous to the species to find ways of utilizing it. In all such cases self-fertilization is not to be thought of as a substitute for broad breeding, but rather as an equivalent of vegetative propagation.

A plant may have the power of vegetative propagation residing in the protoplasm, but may lack the habits of growth necessary to turn this power to practical use. Thus gardeners find it possible to propagate from cuttings many plants which grow in nature only from seeds. The date and various other palms produce, when young, vegetative shoots from the buds of the lower joints, but mature palms no longer produce such shoots. The buds of the trunk bring forth only clusters of flowers. It often happens that of two closely related species one will produce vegetative shoots and the other will not. Thus the Canary Island date palm produces no shoots, though it is a more hardy and vigorous species than the true date palm.

BROAD BREEDING AND LINE BREEDING IN THE SAME SPECIES.

That broad breeding and line breeding both have important uses in nature is shown by the fact that both are assisted by numerous and often highly specialized adaptive characters. These specialized characters become still more significant when we reflect that there appear to be no corresponding adaptations to favor narrow breeding.

Adaptations of plants to secure pollen from abroad and adaptations to insure the use of their own pollen at home both serve a common object in avoiding the worst alternative of narrow breeding. Indeed, the same plants often show both kinds of adaptations at once. The same species may have its flowers adapted to invite far-flying insects or pollen-carrying birds, and may have at the same time devices for excluding ants and other small visitors that can only crawl or fly for short distances and thus bring pollen only from flowers of the same plant or from adjacent plants, which are likely to be closely related.

Many of these apparently contradictory combinations of adaptive characters have been found in nature. They have been used by several writers to show that Darwin's idea of the development of adaptations through natural selection was erroneous, on the ground that it is illogical to suppose that the same plant could at the same time develop adaptations for conflicting purposes.

From the present point of view it becomes evident that there is no biological conflict, but an important physiological agreement between the functions of the two kinds of adaptations. They can be considered as alternative methods of avoiding the same catastrophe of extinction through narrow breeding. Self-fertility can not be of use to a plant without some underlying quality of protoplasmic longevity, like that shown in vegetative propagation, but when that quality exists self-fertilization may take on an importance only second to normal broad breeding. It is not necessary to suppose that self-fertilization takes the place of broad breeding or that it serves all the functions of normal sexual reproduction. A safer judgment is reached by comparing self-fertilization with vegetative propagation, since there is a large measure of agreement in the functions and limitations of the two processes.

That plants with conspicuous or highly specialized corollas are nevertheless capable of self-fertilization does not prove that the corolla is not an adaptive development to aid in cross-fertilization. The specialization of the corolla only shows in a more striking manner that cross-fertilization is of importance to the plant, notwithstanding its ability to propagate by self-fertilization. That a species produces seeds with its own pollen or that it is able to survive many generations of self-fertilization does not prove that the species has ceased to draw any advantage from crossing, even though its opportunities for crossing are relatively rare. The occasional exercise of normal sexuality may be quite as important for a self-fertilizing species as for one whose members are very long lived. Though self-fertilization is a frequent and entirely normal method of propagation for many plants, exclusive dependence upon self-fertilization is a very rare and abnormal condition, doubtless because it leads to ultimate decline and extinction, so that the types which may have adopted this system in the past have not been perpetuated.

LINE BREEDING A SYSTEM OF PROPAGATION.

In domesticated varieties, as in wild species, the chief value of self-fertilization and of other forms of line breeding lies in their use as methods of propagation rather than as true equivalents of sexual reproduction or as means of evolutionary progress. Marked improvements are as little to be expected from self-fertilization as from vegetative propagation. This does not mean that self-fertilized stocks are not worthy of the breeder's close and constant attention, for while there are not likely to be changes in the direction of greater vigor or increased fertility, tendencies in the other directions are frequently manifest. The practical importance of keeping a variety from deterioration may be quite as great as that of causing it to make an actual advance. Differences too slight to be detected in direct comparisons of individual plants of wheat may render one strain distinctly more productive than another, and may make a difference of millions of bushels in the harvest when the superior strain has been spread over the wide regions in which the crop is cultivated.

Unfortunately, this kind of selection, to increase or to maintain the vigor and fertility of a line-bred type, is often the most difficult to carry out in practice, one reason being that these same qualities are most seriously affected by external conditions. All breeders are familiar with the fact that conspicuously vigorous and fertile individuals may yield only ordinary offspring. An individual cotton plant of a Mexican variety grown at San Antonio, Tex., in 1906, produced much more cotton than any other plant in the field, but the offspring of this plant showed no superiority in the next generation.

When individual differences due to differences of conditions are greater than the inherent differences of the plants themselves, the inherent differences become very hard to detect. Even when the comparison is carried over to the progeny by using the test-row or centgener methods, the greatest caution has to be used to avoid mistakes from unrecognized differences of external conditions. The manner in which a field was plowed or fertilized in some previous year may make a notable difference between two rows or plots, may cause the wrong selection to be made, and may thus vitiate an elaborate experiment. The failure of such efforts at the selective improvement of varieties in vigor and productiveness has led some writers to deny that such improvements are possible.

Various attempts have been made to explain the supposed impossibility, some writers going so far as to claim that the characters of the separate lines of descent do not vary at all except as they are affected by differences of external conditions, and hence that all selection is superfluous after " pure lines " of descent have been separated. The fact is, however, that many of the experiments that have been supposed to warrant this conclusion have been carried on for altogether too short a time to justify any reasonable expectation that the results of the selection would become apparent amidst the confusion of fluctuating and environmental variations.

It is to be expected, of course, that attempts to secure a further increase of a particular character must sometimes fail if the limits of expression of the character have been reached or if the particular breed or line of descent has no tendency toward greater variation in that particular direction. But, on the other hand, we may expect with much confidence that a sufficiently persistent selection of vigorous and fertile individuals for propagating a variety will help to maintain these qualities at higher standards than if selection were relaxed and the stock were allowed to become diluted with lines of descent which had lost some of the varietal characteristics by degenerative mutation. Selection of this kind is safer and more likely to insure continuously favorable results when it is not too closely limited to single individuals or to single lines of descent, at least after the initial generation. Unless elaborate precautions are maintained in the testing of each of the lines of descent, a serious mistake may at any time be made by propagating from an inferior individual which has appeared better than its fellows because of some environmental advantage. With a somewhat wider selection there is less danger of discarding the best lines, while those that are inferior are gradually weeded out.

If selection could be finished, once for all, by the separation of a " pure line," any amount of care would be practically justified in making this separation, but if the superiority of a variety has not only to be made by selection, but also to be maintained in the same way, there is less object in narrowing the stock down to the individual line, particularly in the kinds of plants that must be sown or propagated for several generations to obtain enough seed for commercial planting. By the time that the single line has been broadened again to produce commercial quantities of seed, the differences may reappear, so that the type may be no more constant than if several individuals had been taken and a smaller number of generations had been raised. Indeed, our experiments with cotton have shown that uniformity is to be expected with more confidence from a type represented by several closely similar plants than from the progeny of a single individual plant, which is often notably diverse. Thus it may happen, where strict line breeding is impracticable, that a choice must be made between methods that most nearly approach the desired condition.

If selection could compel actual improvements of a stock, vegetative and self-fertilized varieties could be advanced more effectively than any others, for selection can be applied with greater exactness and persistence than in cross-fertilized types. And yet it is in the former classes of varieties that the limitations of selection have been most definitely appreciated.

Even among vegetative individuals of the same stock slight differences are still to be detected, and sometimes quite pronounced changes or " bud mutations " are found. There is no reason to suppose that bud mutations differ in any essential respect from mutations that arise in seedlings of narrow-bred or line-bred varieties. Nor do they need to be thought of as being less sexual than the variations of seedlings, now that we know that the sexual condition (conjugation) continues throughout the existence of the plant even when this existence is prolonged by vegetative propagation. The only distinction that can be made with certainty is that bud mutations are generally much less frequent than seedling variations, there being, apparently, a greater facility of change of characters soon after conjugation begins. Breeders of potatoes believe, according to Dr. E. M. East, that bud mutations are much more frequent in new varieties recently derived from seedlings than in the same varieties after they have been grown for a long series of generations.[a]

Whether the range of diversity in bud mutations is as wide as among seedling variations is hot easy to determine, because the bud mutations are too rare to permit them to be studied in large numbers like seedling variations. Hundreds of seedling mutations of coffee have been observed by the writer in Central America, but only one bud mutation. And yet this showed its divergent characters in a manner quite as pronounced as any of the seedling variations.

The relative infrequency of bud mutations may also be responsible for the fact that they have had little practical importance. Breeders naturally feel that they have much better opportunity of selection among the rich diversities that can be called forth by returning to sexual reproduction. Certain it is that very few valuable new types appear to have been derived from bud mutations—nothing to compare to the superior seedlings of apples, potatoes, sugar canes, strawberries, and many other species. Seldom, if ever, have bud mutations been found stronger or better than the parent stock, except from the standpoint of the florist or fancier interested in multiplying slight differences of form or color.

We use vegetative propagation to preserve the varieties, but resort to sexual reproduction when we wish to improve them, especially in the direction of vigor and fertility. The same vegetative variety may deteriorate faster under some conditions than under others, or it may even regain some of its vigor when conditions are improved. It has been found in Java that sugar canes brought down from elevated localities are more resistant to disease than other representatives of the same vegetative variety which have been grown continuously at low elevations. The change of conditions gives something of the same beneficial effects that are obtained by new conjugations. New seedling varieties of cane have also been found resistant to diseases in the same way as these mountain-grown vegetative strains.

---

[a] East, E. M. A Study of the Factors Influencing the Improvement of the Potato. Bulletin No. 127, University of Illinois, Agricultural Experiment Station, August, 1908.

HOW SELECTION IMPROVES LINE-BRED VARIETIES.

The fact that selection increases the agricultural value of line-bred varieties can be understood without supposing that the plants are changed in any way by the selection. The effect of selection is not biological, but purely mathematical. Selection does not give us any new characters, and does not even raise the vigor or fertility of the plants. It simply gives us the full agricultural use of the particular lines of descent that are showing the vigor, fertility, or other desirable qualities in the highest degree.

A notable improvement of a variety appears to have been made when the best individual strain has been found and the remainder of the variety has been discarded, but this is no warrant for holding that the superior strain is better than it was before. Selection then maintains the superiority of the strain by keeping it from dilution with the lines of descent in which degeneration appears, but still there is no reason to suppose that selection changes the strain.

Though continued selection may add nothing new in the way of an increase of the desirable qualities of the variety, the same reasons for continuing the selection of the variety will always remain that existed before the first selection was made, to eliminate the undesirable diversities which continue to appear even in self-fertilized and vegetative varieties. Careful selection among individuals of such varieties may always be expected to show good results. In old and weak varieties the results will be even more striking than in the stronger and more vigorous stocks, for the greater the tendency to degeneration the greater the contrast will be between the degenerate lines of descent and those that still retain the characters which give the variety its special value.

SELECTION A CONSERVATIVE PROCESS.

Favorable results from individual selections in a narrow-bred or line-bred variety is no proof that the variety as a whole is advancing in excellence, but may indicate quite the contrary—that its vigor is declining. The more rapid the decline the larger is the proportion of degenerate individuals and the greater the practical improvement worked by selection. The greater need of selection in degenerating varieties is like the requirement of more efficient police in degenerate human communities. The removal of the criminals is a practical necessity, though it does not make the community better, except in the relative sense that it may lessen the amount of crime actually committed. In the same way it is necessary to maintain a careful supervision of close-bred varieties to keep them near the standard of efficiency by eliminating the degenerate lines of descent as soon as they fall below the standard.

To compare the best lines of descent with those that have degenerated or with an unselected group containing degenerate lines shows the agricultural value of the selection, but it does not show that the best lines have been made better than they were. Domesticated varieties can be improved agriculturally without being changed biologically. There is no reason to suppose that the sorting out of the degenerate lines of descent in a strictly line-bred variety has any effect upon the lines that are retained. The practical advantage of maintaining selection can be understood without any need of supposing that the best lines are becoming better than they were. If a tendency to degeneration, such as seedlessness, paler color, or softer texture, renders a variety more desirable, the lines which are degenerate, in the biological sense, will be preferred to the others and the variety will appear to be the more improved agriculturally the faster it degenerates. It may be going too far to say that all the instances where persistent selection has resulted in continued improvement in economic characters are examples of biological degeneration, but it is certain that a very large majority are of this nature.

We need to recognize that there are two separate branches of the art of breeding, constructive breeding and conservative breeding, having different objects and requiring different methods. The former attempts to improve the characters of plants and animals; the latter, to preserve and make full use of desirable characters already obtained.

Some would restrict the word breeding to the constructive idea of improvement, which comports with the suggestion that all the processes of line breeding are to be considered as methods of propagation rather than as normal breeding in the sense of sexual reproduction. Nevertheless, such a limited use of the word breeding would exclude a large part of what is now reckoned as breeding, if not, indeed, the whole of the practical part of the subject. Breeders are often very successful in finding new characters not noticed before, as well as in making new combinations of desirable characters and in suppressing undesirable characters, but there is relatively little to support the popular idea that the operations of breeding result in " new creations," in the sense of bringing new characters into existence outside of those already attained in the course of normal evolution of species.

Whatever the differences of opinion regarding the function of selection in constructive breeding, its fundamental importance in conservative breeding will not be disputed. The superiority of the various forms of line breeding over narrow breeding is based on the simple fact that they preserve desirable characters more effectively and for longer periods of time than does narrow breeding. Line breeding is superior to narrow breeding because it is a more efficient method of selection.

The beneficial effects of selection are greatest in self-fertilized varieties because such varieties are no longer in a condition in which they can be injured by selection. A broad-bred variety may be injured or extinguished by selection, because strict and persistent selection puts an end to broad breeding and substitutes narrow breeding in its place. A variety capable of self-fertilization is less injured by narrow breeding, for those lines of descent which continue to be propagated by self-fertilization also continue to escape the bad effects of narrow breeding. Thus the tendency of persistent selection will be to preserve those strains in which the habit of self-fertilization is most strongly developed. When a condition of complete self-fertilization can be reached, as in wheat and barley, there is no longer any question of injury from narrow breeding, since all the lines of descent are propagated by line breeding.

Selection between lines of descent which are already completely separated by self-fertilization makes no further alteration in their conditions or methods of reproduction. Selection then has only the one effect of preserving the lines which express in the most uniform manner the characters most desired. If we take the trouble to choose the best strain we are assured of the best results. In no other way could this assurance be gained, for by any less stringent method of selection there is always the danger that representatives of poor or mediocre stocks are included and that the full possibilities of selection are not attained.

But no matter how effective the methods of line breeding may be in varieties adapted to these forms of propagation, their general indiscriminate application to all kinds of plants is not to be advised, notwithstanding that this course is very frequently advocated, as a consequence of the evolutionary doctrines of Professor De Vries. Rejecting Darwin's doctrine of the evolution of species through gradual changes of characters, De Vries holds that species originate by sudden changes or mutations, like those which appear in our uniform domesticated varieties. Ordinary species and varieties are supposed to consist of mixtures of large numbers of these so-called " elementary species," or " biotypes," originated by mutation. When one of these mutative strains has been separated by selection it is supposed to remain constant and uniform, except as further mutations of individuals may give rise to additional " species."

This doctrine has had in many instances the very beneficial effect of directing the attention of breeders to the advantage of line breeding, but it brings us into conflict with practical facts when it leads us to assume that line breeding is applicable to all kinds of organisms, and also when it teaches that no further selection is necessary after a " pure line " has been separated. No better illustration of pertinent facts could be found than the Triumph variety of cotton, originated by Mr. Alexander Mebane, of Lockhart, Tex., and carefully selected by him for many years. Mr. Mebane's fields of Triumph cotton at Lockhart show wonderful uniformity, as all cotton specialists have agreed who have seen them. And yet when the Triumph cotton is taken to other places, even at no very great distance from Lockhart, many notable deviations from the Triumph characteristics make their appearance. This occurs even in the first generation of plants grown from seed raised at Lockhart, before there is any opportunity for crossing with other varieties or for mutations to take place as the result of conjugation subsequent to the transfer. The Triumph embryos in the seeds or the young plants after they have germinated are able to change their characters and depart from the Triumph type. These departures, moreover, are commonly inherited by all the progeny of the variant individuals. The persistence of such variations soon puts an end to the uniformity of a variety unless a vigilant selection be maintained.

UNIFORMITY NOT A NORMAL CONDITION OF HEREDITY.

That the effects of selection upon self-fertilized varieties are always agriculturally beneficial makes it possible to understand the growth of the idea that uniformity is a fundamental principle of heredity and the first ideal of the breeder's art. Uniformity has come to be considered as an object in itself, for the very practical reason that selection for uniformity tends to carry self-fertilized varieties over from narrow breeding to line breeding even when the breeder has had no intention to make such a change of method. To adopt as a fixed ideal of selection the characteristics of the best individual member of the variety generally amounts in practice to the saving of the progeny and nearest relatives of the superior individual. The more strict the adherence to the ideal, the greater the approximation to line breeding and the better the results.

Nevertheless, it must be admitted that the value of uniformity as a method of breeding one class of varieties does not establish it as a general principle of heredity, to be applied to all varieties alike. Nor does the behavior of self-fertilized domesticated varieties show that uniformity is the natural condition of species in nature. Closer examination of the facts enables us to see why uniformity is desirable in agricultural varieties, and also why selection for uniformity tends to produce the desired result. We do not find that the idea of uniformity applies to wild species nor in the human family, nor even in the dogs and other domestic species which we learn to know as individuals. Nor have we any assurance that uniformity is a safe general guide in the field of agricultural breeding. Even in

146

dealing with line-bred types we are no longer satisfied with likeness between individuals, but have learned to separate them into lines of descent and to judge of the value of a line of descent by the general quality of successive generations of progeny instead of from single individuals. In broad-bred types uniformity is obviously not the ideal, except to the extent that it may be necessary for commercial and industrial purposes.

TWO FORMS OF MASS SELECTION.

Writers on evolution have drawn distinctions between several different forms or conditions of selection. The selection which goes on in nature, without human interference, Darwin called "natural selection." Of selection by man Darwin distinguished two forms, unconscious selection and methodical selection. Selection becomes conscious when it has for its object the improvement of the stock. The selections carried on by primitive peoples are usually quite as unconscious as those made by birds, insects, or inanimate conditions of the environment. Thus, the Indians of eastern Guatemala have a very early, quick-maturing type of cotton, the development of which has been assisted, no doubt, by the fact that they begin to gather the cotton as soon as the bolls begin to open. This has insured the saving of seed from the earliest plants, not because of any idea of improving the crop in earliness, but because the dry season of that part of the country is short and uncertain. If the rains come too early, only the first pickings are saved; late-maturing plants are completely excluded from the stock.

The practical difference between this unconscious selection for earliness and the conscious selection practiced in more intelligent communities lies only in the keeping of the earliest pickings separate from the others. Among the Indians all the pickings are likely to be mixed together in favorable seasons, so that a definite selection takes place only in unfavorable years.

Later writers go beyond Darwin and divide methodical selection again into two forms, commonly called "mass selection" and "individual selection." This is because the progeny of single individuals kept by themselves often maintain higher averages than the progeny of several superior individuals of the same strain when allowed to interbreed. From such facts the inference has been drawn that mass selection is essentially inferior to individual selection and should everywhere be abandoned. Some writers even go so far as to deny that any improvement can be wrought by mass selection, while others object that it is slow and inefficient. Nevertheless, practical breeders are loath to abandon as worthless the method by which all the earlier

146

improvements of our seed-propagated plants were made—the improvements which have carried them so far beyond their wild relatives in the special qualities and which render them useful to mankind.

In reality, two essentially different conditions are being confused under the name mass selection, the conditions which have been distinguished as broad breeding and narrow breeding. When mass selection is applied to a carefully selected variety or strain, derived, perhaps, from a single ancestor, the condition is the same as that here described as narrow breeding. When mass selection is applied to larger groups with more normally diverse ancestry, we have a condition of consciously directed broad breeding, a form of mass selection which may be highly beneficial.

The objections commonly urged against mass selection apply to the former condition, but not to the latter. Mass selection in a self-fertilized plant like wheat is inferior to individual selection, or line breeding, simply because it is less thorough, as already explained in previous chapters. Mass selection gives us a collection of the better of the individual strains, which must always be found inferior to the best of the strains whenever it is practicable to separate these from the others. In cross-fertilized plants and animals mass selection may be inferior for the further reason that the narrow individual crosses arouse undesirable degenerative variation.

In its application to self-fertilized plants a persistent individual selection constitutes line breeding. Nevertheless, it is dangerous to prescribe individual selection as a solution for all the problems of breeding, for it solves problems of only one kind and gives wrong indications in the others. The theory of individual selection neglects the practical distinctions between the different forms of line breeding as well as the differences between narrow breeding and broad breeding. The expression "individual selection" is in itself rather misleading, for all the methods of selection involve the selection of individuals. The difference of methods does not lie so much in the selection of individuals as in the breeding they receive after they have been selected.

THE "RUNNING OUT" OF VARIETIES.

Some horticulturists believe in the "running out" of varieties of apples, pears, and other tree fruits propagated from grafts or cuttings, while others have been unwilling to admit that varieties run out. Such differences of opinion are possible in dealing with the varieties of long-lived trees, for our varieties have not been known far beyond the span of life possible for individual trees under favorable conditions. With potatoes, strawberries, lilies, carnations, and sugar cane the weakening of old varieties is a commonly recognized fact. Specialists in potatoes and strawberries estimate the productive life of a variety between twenty and sixty years. It is the opinion of Dr. B. T. Galloway that the frequent substitution of varieties among growers of carnations is due as much or more to the greater vigor of the new seedlings than to any superior attractiveness of the flowers. The commercial life of a variety of carnations seldom lasts more than three or four years.

As vegetative varieties of many different kinds have shown the same tendency to become sterile, so it is possible that carefully selected types may reach a condition where any admixture of blood brings deterioration, as breeders sometimes claim. Improvement by crossing would then be precluded. The particular character which affords our standard of the excellence of a variety may appear for a time to increase under selection, but its highest expression is finally reached. Sterility or other forms of weakness become more and more apparent. New types arise to claim the superiority, and the old are soon abandoned and forgotten. To know that a method of breeding leads ultimately to degeneration may not remove it from the field of agricultural utility if there are compensating advantages. It may be easier to secure new varieties to replace the old ones as they decline than to apply any effective methods for avoiding the decline.

Those who are concerned to prevent the extinction of species of wild animals and plants may find an object of even greater practical interest in the danger of exterminating the wild types of our domesticated species, which future generations may find indispensable for replenishing the stock of domesticated varieties.

Though the limits can not be determined without making the experiment, we must not expect that selection or any other means can lead to an unlimited expansion of any character. Cotton plants can not be turned completely into lint, nor beets made to yield 100 per cent of sugar. Characters should not be thought of as independent entities capable of indefinite expansion; in reality, characters are only parts or functions of highly coordinated complex organisms. The existence of each character represents a cooperative result of the activities of other parts, so that the whole organism has to be unbalanced in order to permit an excessive expression of a particular part.

REJUVENESCENCE OF VARIETIES.

It is not to be assumed that all of our carefully selected varieties have reached a condition where crossing can work no improvement. For many of them it may still be of the greatest importance to provide normal fertilization at needed intervals.

The object of such crosses is not that of changing the characters of the stock, but to restore the vigor. Some breeders have reported excellent results from the rejuvenescence of varieties by crossing, but others have drawn back from such attempts because they encountered an unwelcome amount of variation. That diversity should appear as the result of crosses between varieties is of course to be expected, but it may be possible to eliminate this diversity in a few generations and yet retain an increased vigor for many generations.

In normal species, as we have seen, line breeding is not a full equivalent of sexual reproduction, but a supplementary system of propagation. Nature shows that broad breeding and line breeding may be combined, but shows also that they should not be confused or compromised into the injurious intermediate condition of narrow breeding. In domestication, too, it may often be found advantageous to combine the two methods, without confusing them. We do not insist upon growing all potatoes from seed because new seedling varieties are often superior to the older stocks which have been subjected to many years of vegetative propagation. In the same way we may find it as advantageous to resort to occasional crosses in seed-propagated plants as to adopt newer and stronger seedling varieties in species adapted to vegetative propagation.

Crossing between sufficiently remote strains of a line-bred variety may also be found highly advantageous and may help to maintain vigor, notwithstanding the fact that crosses between more nearly related lines of descent may yield results inferior to those of strict line breeding.

Even without crossing it is sometimes possible to secure a notable increase of vigor or other desirable quality through the discovery of an unusual individual. Mutations often show the same sudden increase of vigor as hybrids, and this may continue in their descendants for many generations. Many important varieties of plants and breeds of animals trace their origin and popularity to a single superior individual. Though we can hardly say that we rejuvenate an old breed when we replace it with a new one, the same agricultural purpose is attained in the two cases, especially if the agricultural qualities are the same.

A BALANCE BETWEEN BROAD BREEDING AND NARROW BREEDING.

A system which has been followed with apparent advantage in some of the carefully bred varieties of domestic animals is that of wide and continuous crossing inside the breed. Instead of limiting descent to a single individual strain or to a few related strains, the effort is made to breed together individuals which show all the characters of the breed and yet have as little consanguinity as possible. Such a system, if efficiently carried out, as among the Australian

146

sheep breeders, keeps the lines of descent of the variety connected in a network. Though the network is narrower than in a natural species it is at the same time much broader than in ordinary narrow breeding.

Such a condition might be described as the narrowest form of broad breeding or as the broadest form of narrow breeding; it represents the boundary or balance between the two methods. It is the condition in which a considerable degree of uniformity is maintained and deterioration from narrow breeding is avoided as far as possible. For convenience of expression it seems best to consider this condition as one side of the field of broad breeding, because use is being made of crossing between different lines of descent to maintain the vigor of the stock. The practical point is to avoid the condition of still narrower breeding between this balanced condition in which the physiological effects of crossing are still utilized and the condition of definite line breeding in which narrow crossing is excluded and reliance is placed on the other principle of protoplasmic longevity.

AVOIDANCE OF UNNECESSARY UNIFORMITY.

How to maintain fertility by broader breeding and at the same time avoid an undesirable diversity of characters is one of the practical problems which each breeder must consider from the standpoint of the types with which he deals. By proper attention to the expression relations of the different characters it may be possible to secure diversity of descent with relatively little diversity of expression or to confine diversities of expression to characters which will not interfere with the utility of the breed.

In the cotton plant, for example, many characters might be allowed to vary freely without interfering in the least with the value of the product, and such variations would have positive agricultural value if they gave increased vigor and fertility. If the plants produce early crops of large bolls with lint of uniform length and fineness it makes no difference that stems have different colors or different amounts of hairs, that the leaves are differently cut, the flowers differently spotted, the nectaries and bractlets differently shaped, or the oil glands differently distributed.

This question of the amount of uniformity to be required is of primary importance in the domestication or acclimatization of new types of plants, and especially of field crops that are subject to cross-fertilization, like cotton and corn, where strict line breeding is not practicable. Not to require unnecessary uniformity in such cases does not mean that selection can be safely relaxed in types that have already been brought to a condition of uniformity, nor does it justify us in overlooking the fact that characters of no value in themselves may be worthy of careful attention when they are correlated with more important features. With cotton, for example, there seems to be a very definite correlation between the external form of the boll and the length of the lint. Plants that have longer and more pointed bolls are almost always found to have longer lint than neighboring individuals with more rounded bolls. It is therefore worth while to pay attention to the shape of the boll as well as to its size if the desirable quality of long lint is to be carefully guarded.

Similar correlations between unimportant characters and those that have definite utility have been detected by Mr. W. W. Tracy, sr., in varieties of corn, beans, lettuce, and other plants. In the Extra Early Adams variety, corn plants with branched tassels, even though they may be as early themselves as other plants with the unbranched tassels characteristic of this variety, have been found to yield offspring with a distinct tendency to later maturity. Plants of the Golden Wax beans with flowers slightly larger than normal were found to yield offspring with a distinct tendency to produce ordinary green pods instead of yellow. Certain varieties of lettuce which differ in the characters of the cotyledons have also been found to differ in the time of running to seed, even when they appear to be indistinguishably alike in the interval between.

The corn plant has been the object of much selection, but this has been directed almost exclusively to the characters of the fruit instead of to the plant as a whole. The result is that even in varieties that give relatively uniform ears and kernels the vegetative characters continue to show wide ranges of diversity in form, color, and hairiness, as well as in the tassels and flowers.

In a broad-bred stock it is evidently possible for many such features to continue their natural diversities without detriment. It is only in narrow-bred types that absolute uniformity has appeared a desirability of breeding, and this we can now understand, because the application of such standards has always tended to carry these types toward the superior condition of line breeding. The efforts that have been made to place cross-fertilized types, like Indian corn, on the same basis of uniformity as a line-bred type, like wheat, have not had any corresponding measures of success. In spite of long and persistent selection only a small proportion of the plants show any complete uniformity of the fruiting characters. The facts with reference to corn have been summarized by Mr. A. D. Shamel of this Department.

It is the experience of most corn breeders that it is not possible to produce on an acre more than 5 bushels of uniform ears even of our most improved strains. A large majority of the plants produce ears of small size, irregular

146

shape, and light-weight, which are undesirable. Many of the stalks are barren. Only a small proportion of the plants produce the maximum size and weight of ear.[a]

Such facts make it evident that there is a very practical distinction between the cross-bred and the self-fertilized types with reference to uniformity. Selection toward uniformity increases the yield of self-fertilized types, because their methods of reproduction are well adapted to give regularity in the expression of the desired characters, whereas the cross-fertilized types have not the same power of producing uniform offspring. With the cross-fertilized plant there is no such direct and necessary connection between uniformity in high yield and uniformity in other characteristics. If corn could be propagated by self-fertilization it might be expected to respond to selection in the same way as the self-fertilized cereals, but no self-fertilized strains of corn have been developed. In the great majority of corn varieties self-fertilization is definitely prevented by the habit of proterandry, the ripening of the pollen of the plant before its silks are ready for fertilization. In an experiment with many varieties at Lanham, Md., in the season of 1908, a single variety proved to be an exception to this rule. The stalks produced silks and pollen at the same time, but the variety can hardly be said to be in a normal physiological condition, for the leaves of all the plants were thickly spotted with yellow, a symptom not shown in any other variety.

Several experimenters have reported distinctly unfavorable results from self-fertilization in corn, the self-fertilized progeny being notably inferior to the cross-fertilized. It is possible that strains of corn may be found which will thrive under self-fertilization, but this is rendered somewhat less likely by the fact that the stamens and pistils are produced on different parts of the plant, instead of close together.

The internodes of the higher plants have, as botanists know, a certain individuality of their own, the plant being a complex or colony of many of these internode individuals. If we consider the plants from this standpoint of the individuality of the internodes it becomes plain that the pistil and stamen internodes are very close relatives in the self-fertilized cereals, while in the corn they are separated by many generations of internodes, so that their relationship is relatively remote. The self-fertilization of a corn plant does not mean the uniting of cells from adjacent internodes, as on the wheat plant, but unites cells which are in some senses as little related to each other as though they were on different plants.

---

[a] Shamel, A. D. The Art of Seed Selection and Breeding. Yearbook of the Department of Agriculture for 1907. p. 227.

Thus it appears that the specialized habit of corn to produce its pistils and stamens on different parts of the plant puts it out of the reach of the kind of self-fertilization practiced in wheat. In order to make corn experiments which shall be truly parallel with those of the self-fertilized cereals, it will be necessary to find varieties with hermaphrodite flowers, indications of which are sometimes found on abnormal suckers of our ordinary varieties.

In view of these unusual obstacles to line breeding it is fortunate that corn has relatively little practical need of the uniformity which is of so much importance in crops like cotton and tobacco. The chief object in a corn crop is a large yield. A large proportion of the corn is not sold at all, but is used on the farms where it is raised. Farmers would be very willing to permit variations, even in the shapes and colors of the grains, if they could get enough more corn per acre.

Experiments have shown that increased yields can be obtained by crossing varieties of corn. Indeed, the fact seems to have been utilized since prehistoric times among the Indians of the Quezaltenango region of western Guatemala. They follow the practice of planting three different types together for the bulk of their crop, in the belief that larger yields are obtained in this way than when the varieties are planted separately. Nevertheless, breeders remain loath to advise mixed plantings of corn, or even to investigate such a possibility of increasing the yield, perhaps because this advice is so far out of accord with the methods by which some other cereal crops have been improved.

THE " FIXING OF CHARACTERS " BY LINE BREEDING.

Recognition of the superiority of line breeding over narrow breeding may help to explain the belief of many breeders that inbreeding "fixes" characters. Characters may appear to be fixed because line breeding protects a selected stock from the degenerative diversity which comes with narrow breeding. When such a line-bred stock is crossed with a narrow-bred variety there may be a definite tendency for the peculiarities of the superior stock to predominate in the offspring.

The mutations of narrow-bred and self-fertilized plants show that mere repetition does not insure stability, but may be followed by wide diversity, even after uniformity has been maintained for many generations. Selection has been supposed to eliminate undesirable characters, but in reality it is only able to postpone changes of expression relations. The characters which the breeder would eliminate continue to be transmitted and may continue to reappear, even in the face of persistent inbreeding.

146

RECURRENCE OF DIVERSITY IN CROSSES BETWEEN LINE-BRED GROUPS.

Further evidence that self-fertilization and in-and-in breeding are to be associated with vegetative propagation rather than with normal conjugation is found in the fact that diversity of expression reappears when diversity of descent is permitted to members of line-bred varieties.

Normal broad breeding is accompanied by alternative expressions of characters involving continual readjustments of the internal relations which govern expression. Line breeding, on the other hand, avoids such readjustments. With respect to the expression of its characters a whole line-bred variety corresponds to a single individual of a broad-bred group. Differences between line-bred varieties correspond to differences between the individual members of species, not to differences between the species themselves. All the forms of line breeding yield a relatively great uniformity of characters, but there are suggestive differences in the readiness with which the latent diversity reappears.

The uniformity secured by vegetative propagation conceals but does not diminish the inherent diversity of expression relations. The individual diversity of seedlings of vegetative varieties is familiar to all propagators of such plants. Self-fertilized types may also show a ready recurrence of diversity of characters when sufficient diversity of descent is supplied to correspond to that of normal interbreeding among the diverse members of a species. In strictly self-fertilized types, like wheat, crossing between different individuals of the same variety may bring out diversities as wide as those shown by crosses between different varieties.

In plants having a normal aptitude for self-fertilization the uniformity of varieties is to be thought of as due to this fact of self-fertilization rather than to selection alone. In a strictly self-fertilized type, selection can have reference only to the efficiency with which the characters are maintained under self-fertilization. It appears to have no effect at all upon what may happen when crosses are made, even between individuals of the same variety. The adjustment which suffices for uniformity under self-fertilization would not be sufficient to maintain uniformity under in-and-in breeding.

The effect of persistent selection is to conserve those lines of descent which have great uniformity of expressing relations along desired lines, for the others are rejected whenever their expression deviates in any appreciable manner from the standard set by the breeder. Hence the desirability of definitely directing our selection toward uniformity of expression, as in the centgener method, which tests the different lines of descent with reference to the quality of uniform excellence of the progeny rather than with reference to exceptional attainment on the part of particular individuals among the progeny.

In types not definitely restricted to self-fertilization selection for uniformity has a more definite influence upon the results of conjugation. With such types diversity comes less readily to the surface, even with crossing, but is still able to show itself in striking forms of mutative variation. Hybrids between two related varieties of Upland cotton commonly show a much smaller range of diversity than the mutations which appear in the pure-bred parental stocks. The diversity between varieties is not great enough to overcome the effects of selection and arouse the latent diversities which appear in the mutations, but when wider crosses are made between unrelated stocks the hybrids offer a wide range of diversities, corresponding to the diversities among the mutations.

CONCLUSIONS.

Long-standing differences of opinion among breeders regarding the values of crossing and inbreeding can be reconciled by recognizing the fact that there are three primary conditions or methods of reproduction, instead of two. The three conditions may be called broad breeding, narrow breeding, and line breeding.

In broad breeding there is no restriction of descent to particular lines. Unions are freely made among large numbers of lines of descent, as in natural species, where lines of descent are united into broad, continuous networks. In narrow breeding descent is restricted to unions among only a few lines, forming a narrow network of descent. In line breeding descent is restricted to simple lines, so that no network of descent is formed.

Line breeding is superior to narrow breeding, but can not be considered superior to broad breeding except for special purposes of commercial production requiring a high degree of uniformity. Broad breeding is the condition of normal evolution of species; narrow breeding the condition in which degeneration most promptly takes place; and line breeding the more stable and uniform condition desired in many domesticated animals and plants. Broad breeding is constructive, narrow breeding destructive, and line breeding conservative.

The uniformity of a group is increased by restricting descent so that all the individuals are produced from few ancestors instead of from many. Descent from many diverse ancestors maintains the individual diversity of natural species. Diversity lessens as descent is restricted, but there is also a gradual decline in the vigor and

146

fertility of the stock. The practical importance of uniformity should not cause us to overlook the fact that uniformity must be attained at the price of deterioration. We must either avoid the deterioration of varieties or replace them frequently with new varieties.

Though all forms of restricted descent lead ultimately to degeneration, the decline may be exceedingly slow and gradual if methods of line breeding are followed. Line breeding has practical superiority over narrow breeding when it preserves desirable strains of plants or animals for longer periods of time. The ability of some species and varieties to maintain themselves under line breeding is to be considered as a form of longevity, depending on the power of the protoplasm to continue its activity without new conjugations between germ cells from different lines of descent. Species and varieties differ in their ability to persist under restricted descent, just as they differ in the longevity of the individual organisms.

Four forms of line breeding may be distinguished: Vegetative propagation, parthenogenesis, self-fertilization, and in-and-in breeding. All stages and gradations can be found, from the broad breeding of natural species, through the narrow breeding of ordinary domesticated varieties, to the strict line breeding of individual strains. The superiority of broad breeding over narrow breeding depends on the factor of normal conjugation between cells derived from different lines of descent, whereas conjugations between cells of too closely related lines often produce weak or abnormal offspring.

If conjugations are to take place between unlike germ cells, a considerable degree of diversity of parentage is to be maintained. If conjugations are to be limited to closely related germ cells, diversity of parentage is to be avoided. In broad breeding we imitate the methods of descent in natural sexual species. In line breeding we follow the analogy of vegetative propagation and self-fertilization. A truer idea of the value of other methods of line breeding is gained when we associate them with vegetative propagation than when we consider them as equivalents of the sexual reproduction of broad breeding. That no group of higher plants or animals relies for its reproduction upon any form of restricted descent forbids the assumption that domesticated varieties can be permanently maintained under conditions of restricted descent.

The superiority of vegetative propagation shows that complete cessation of conjugation preserves varieties better than conjugation without diversity of descent. In parthenogenesis conjugation is likewise in abeyance, the vegetative growth of a new organism taking the place of the formation of new sex cells. Self-fertilization is not far removed from parthenogenesis, for the cells that unite have only recently separated from the same parent and are able to combine without disturbing the expression relations of the parental characters.

Evidence of the superiority of line breeding over narrow breeding is also found in nature in the adaptations by which many wild species avoid narrow breeding. Such adaptations are not to be considered as opposed to broad breeding, because of the many cases where the same flower has two kinds of adaptations, some favorable to broad breeding and others to self-fertilization, but both tending to prevent narrow breeding. Thus it appears that vegetative propagation and other natural forms of line breeding are to be considered as supplementary to broad breeding rather than as substitutes for broad breeding.

The superiority of line breeding over narrow breeding depends on the factor of longevity in the protoplasm, to enable the growth of new individuals to continue without the need of frequent recourse to the physiological stimulus of conjugation. Varieties having the necessary longevity are propagated more successfully by line breeding than by narrow breeding, showing that protoplasmic longevity is able to sustain the vitality of such stocks better than the conjugations which occur under conditions of narrow breeding without adequate diversity of descent.

The superiority of line breeding over narrow breeding explains the improvement often wrought by closer selection in narrow-bred groups. Varieties may be improved by more rigid selection for uniformity if they are thus carried from the condition of narrow breeding toward the more favorable condition of line breeding. Varieties may be injured by more rigid selection when it carries them from broad breeding to narrow breeding, especially those varieties which can not be placed on a basis of line breeding for lack of the necessary longevity.

The effect of line breeding is to restrict the expression of characters to a single individual set by suppressing the original diversity of the group. Nevertheless, the suppressed characters have a persistent tendency to return to expression, especially when opportunities are afforded by crossings between different lines of descent or by changes in external conditions. There is no warrant for believing that any method of selection can establish varieties on a stable basis so as to prevent the return of diversity and render further selection unnecessary. Selection always appears to improve narrow-bred and line-bred varieties, not because it raises them to new standards but because it weeds out those lines of descent which have failed to maintain the old standards.

The fact that different physiological principles are involved in the different methods of reproduction practiced in the various kinds of

146

plants and animals shows that no single generalization can be applied to the whole field of practical breeding. The success of line breeding in some cases does not warrant the advice that line breeding be applied to all cases, nor do particular failures with line breeding justify any general insistence that crossing must be practiced in all varieties. The practical need is to recognize the effects that the different methods of breeding are exerting upon our varieties, so that we may guard them against deterioration as long as possible and provide other varieties to replace them.

146

INDEX.

146

O

•